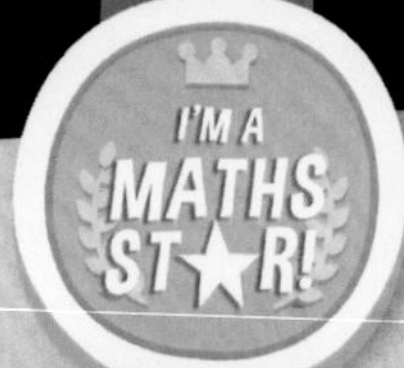

The Boy
and
the Ants

by Dr Boaz

Illustrated by
Nicodemus Loh

Includes a Guide to
Problem-Solving by Yeap Ban Har

WS Education

NEW JERSEY · LONDON · SINGAPORE · BEIJING · SHANGHAI · HONG KONG · TAIPEI · CHENNAI · TOKYO

Published by
WS Education, an imprint of
World Scientific Publishing Co. Pte. Ltd.
5 Toh Tuck Link, Singapore 596224
USA office: 27 Warren Street, Suite 401-402, Hackensack, NJ 07601
UK office: 57 Shelton Street, Covent Garden, London WC2H 9HE

National Library Board, Singapore Cataloguing in Publication Data
Name(s): Boaz, Dr. | Loh, Nicodemus, illustrator.
Title: The boy and the ants / by Dr Boaz ; illustrated by Nicodemus Loh.
Other Title(s): I'm a math star.
Description: Singapore : WS Education, [2022] |
         Includes a guide to problem-solving by Yeap Ban Har.
Identifier(s): ISBN 978-981-12-5052-1 (hardcover) |
         978-981-12-5110-8 (paperback) | 978-981-12-5053-8 (ebook of institutions) |
         978-981-12-5054-5 (ebook of individuals)
Subject(s): LCSH: Mathematical recreations. | Mathematics.
Classification: DDC 793.74--dc23

British Library Cataloguing-in-Publication Data
A catalogue record for this book is available from the British Library.

For any available supplementary material, please visit
https://www.worldscientific.com/worldscibooks/10.1142/12674#t=suppl

Desk Editor: Daniele Lee
Design: Eliz Ong

Printed in Singapore

It was a nice summer's day.

The boy laid contentedly on the grass

and saw some ants on a stick.

The stick was four metres long.

4 metres

There were one, two, three, four, five, six,
seven ants

crawling on the stick.
They were moving along the
length of the stick.
But how strangely they moved ...

An ant would crawl towards the right

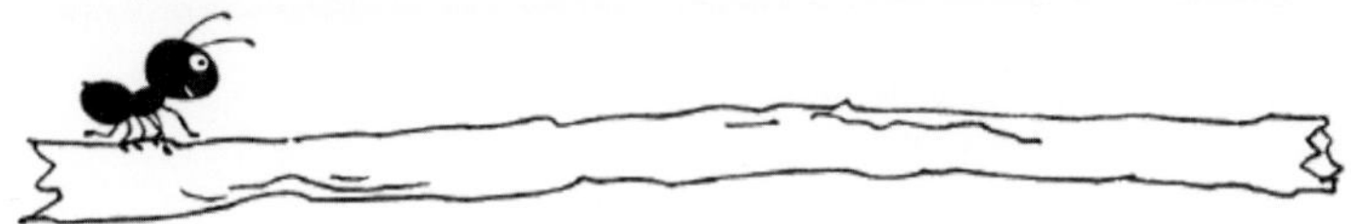

and crawl on

until it met another ant crawling left.

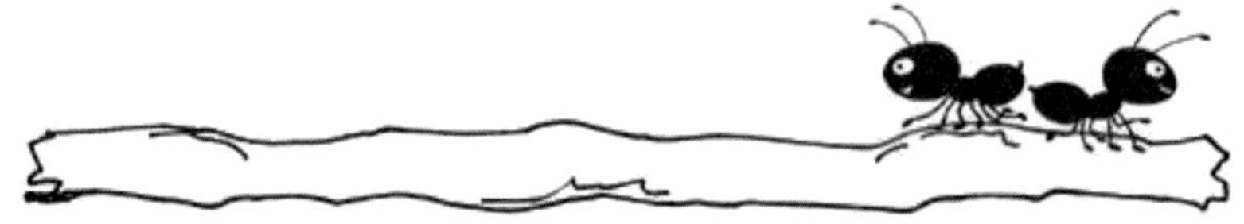

Then both ants would turn around

and move back to where each came from!

The same was true for all the ants.
Left or right, each would move

until it met another moving the opposite way.

Hello! Goodbye! Turn and go!

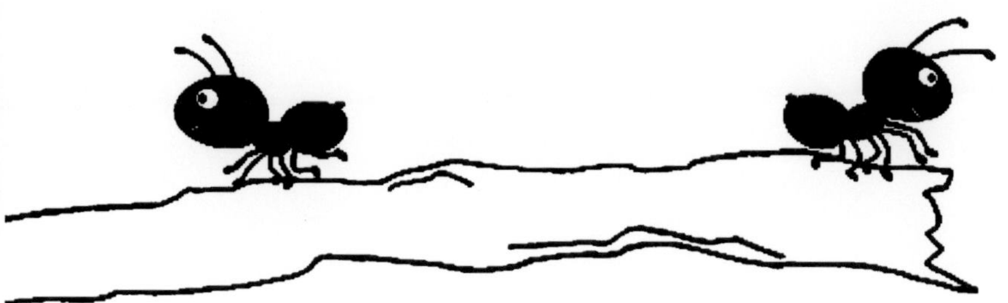

Now, if one reached the stick's end,

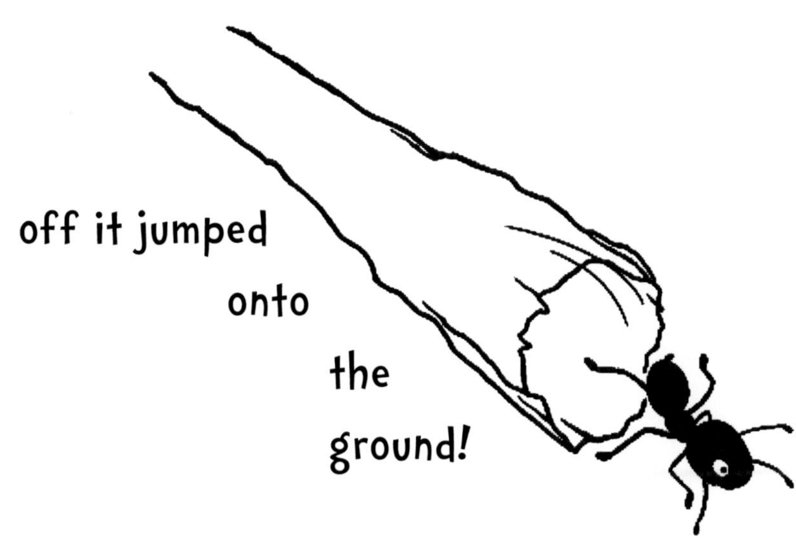

off it jumped

onto

the

ground!

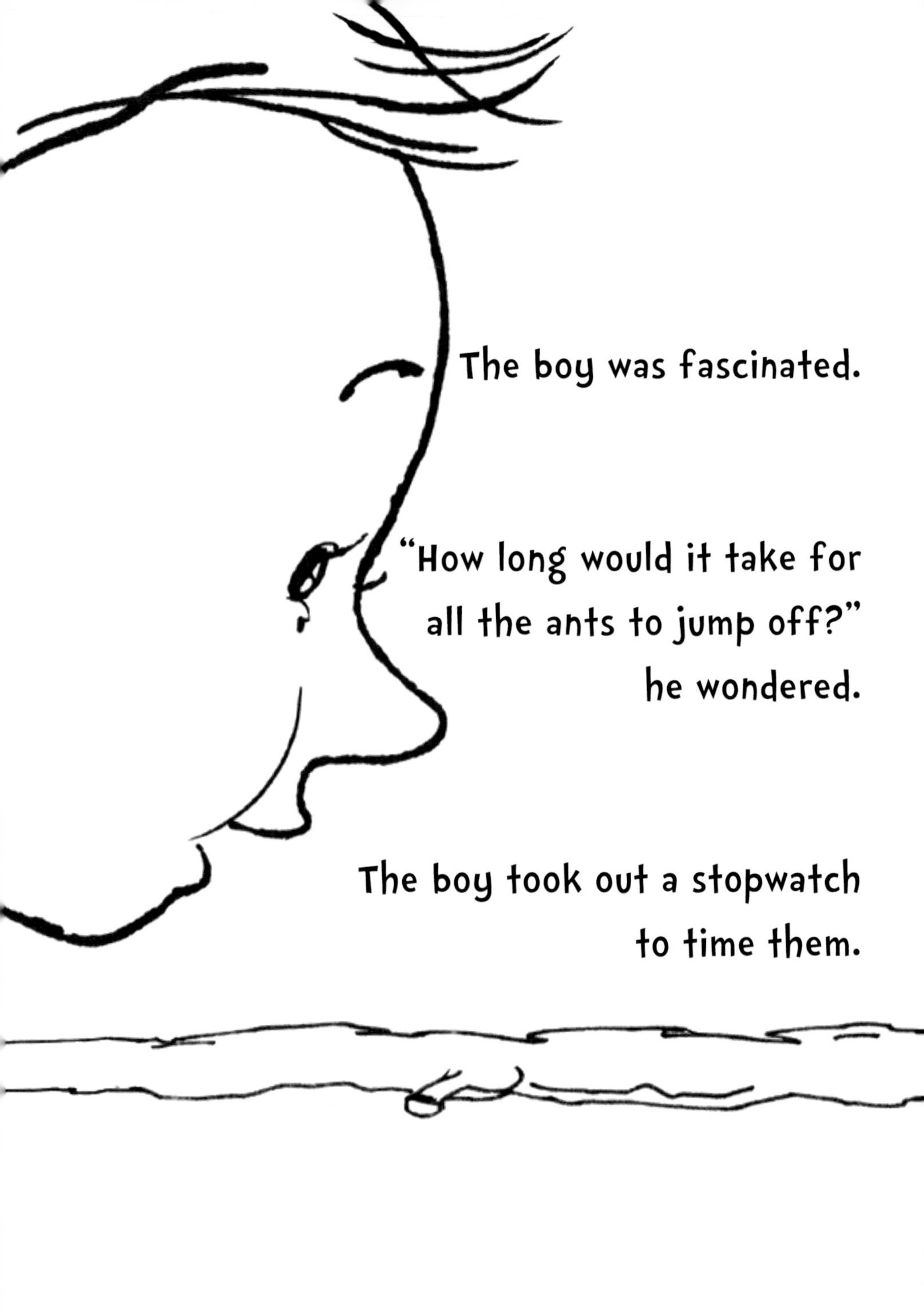

The boy was fascinated.

"How long would it take for
all the ants to jump off?"
he wondered.

The boy took out a stopwatch
to time them.

But he was too late.
All the ants had jumped off already!

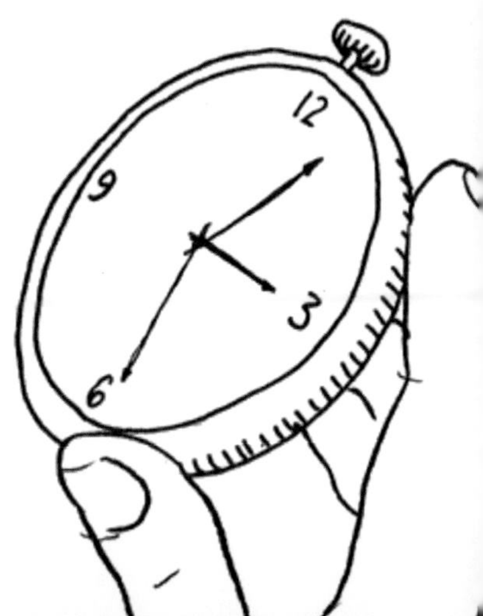

"I'll find another stick," the boy said.

And yes, there was another stick,
four metres long and with seven ants on it!

Back and forth they moved.
Hello! Goodbye! Turn and go!

Quickly, the boy took out his watch.
He could see that the ants moved

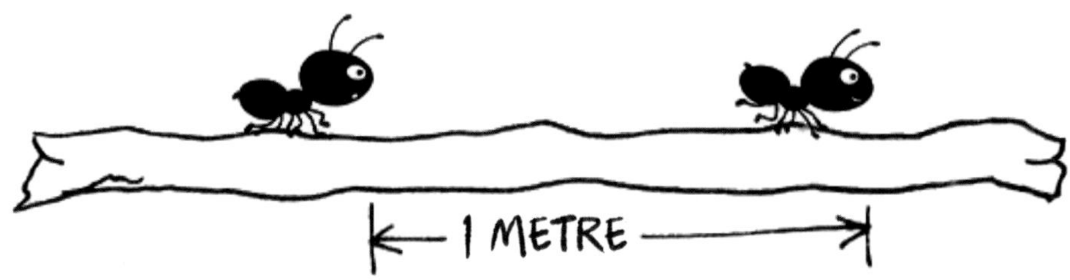

one metre in fifteen seconds

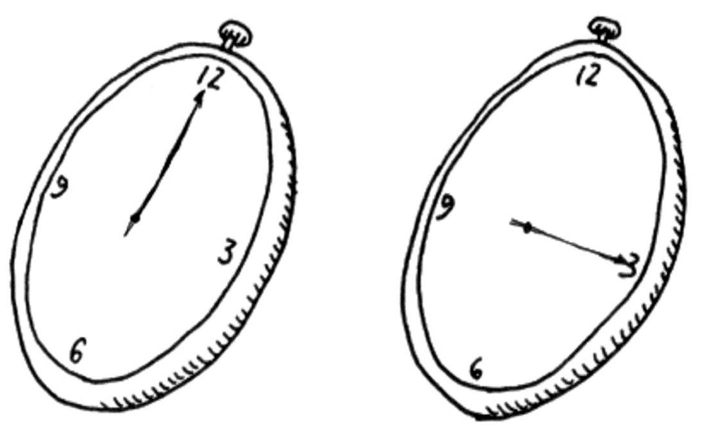

— that meant four metres in a minute.

All the ants were marching at the same pace!

The ants bumped.
The ants turned.

The ants bumped again.
The ants turned again.

And the boy was confused.
It was hard to keep track
of all the ants!

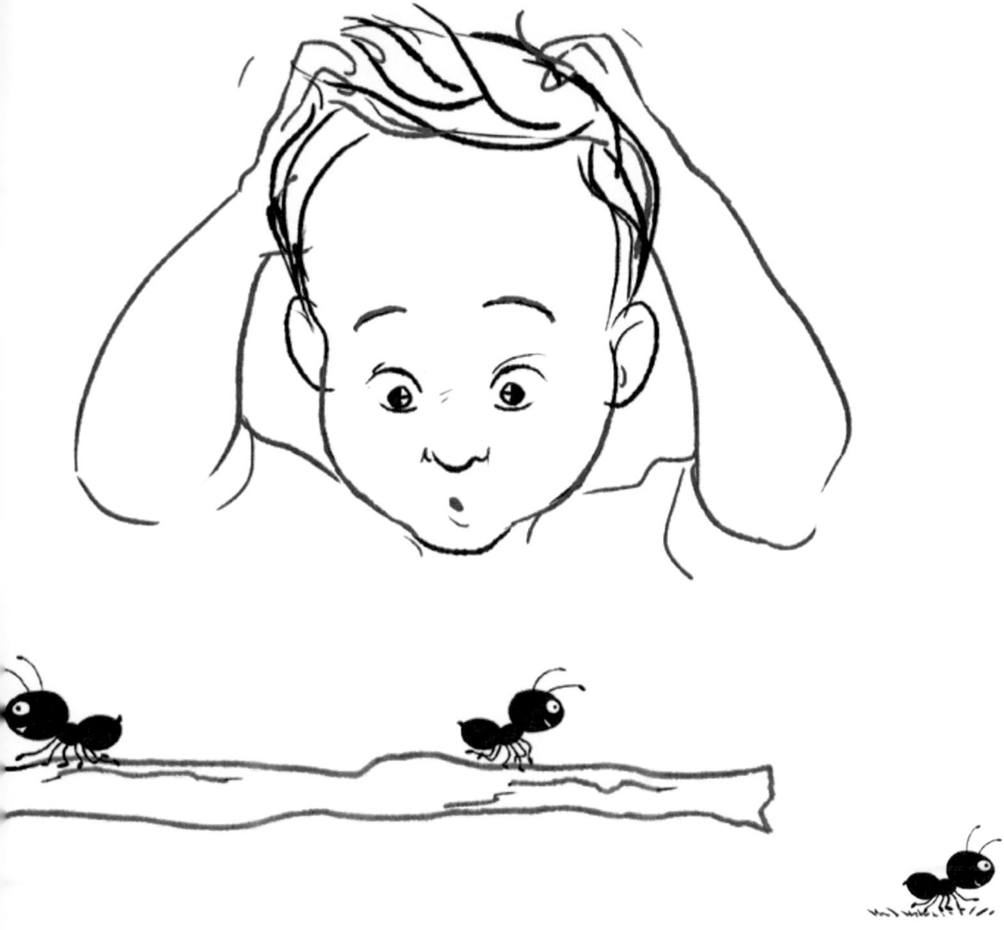

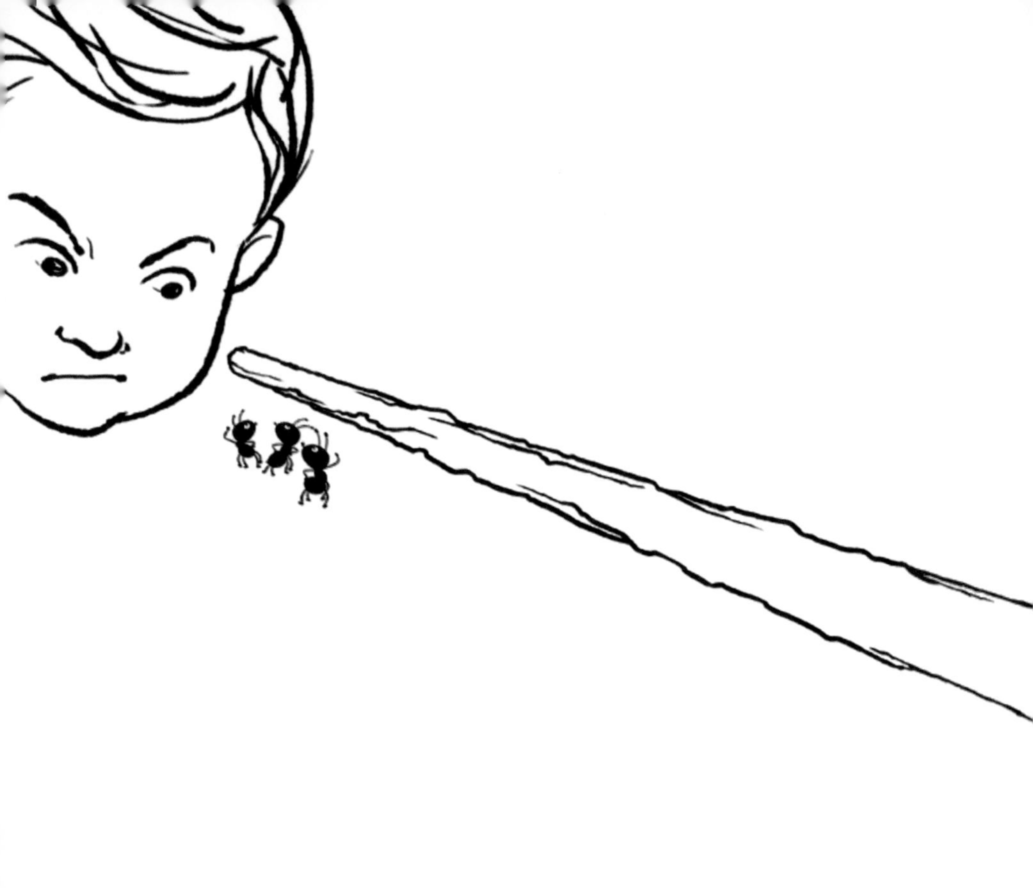

Now all of them had
jumped off again!

The sun was high in the sky

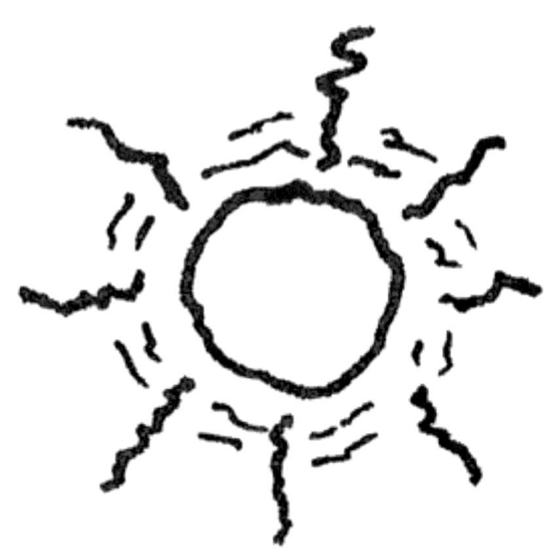

and it was warm and pleasant.

The boy closed his eyes.

"Let me imagine something simpler," he said.

"If there were only one ant,

it would crawl in one direction till the end.

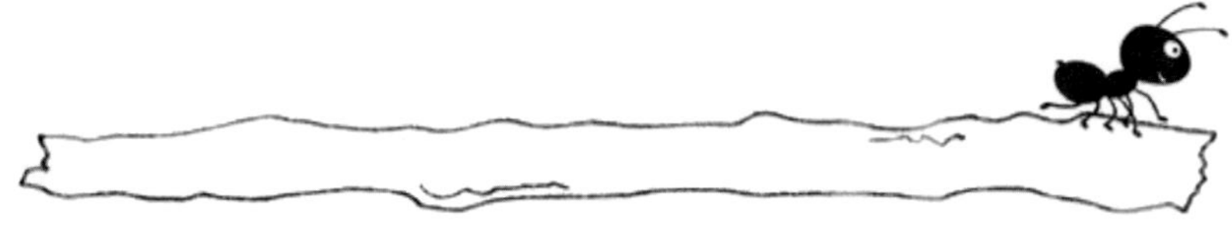

If there were one ant,
and it started from one end
and walked four metres to the other end,

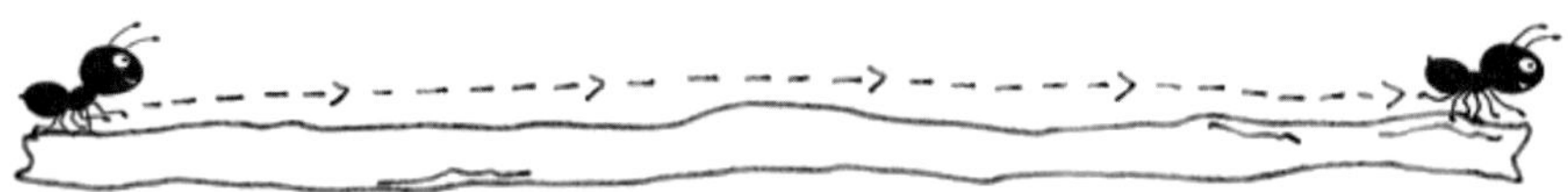

it would take that ant one minute.

If the ant didn't start from one end, but
started from somewhere along the stick,
then it would take not more than one minute.

If there were two ants,
one moving right from the left edge
and one moving left from the centre,

they would meet at the one-metre mark, and
1/4 of a minute would have passed.

The left ant would jump off soon

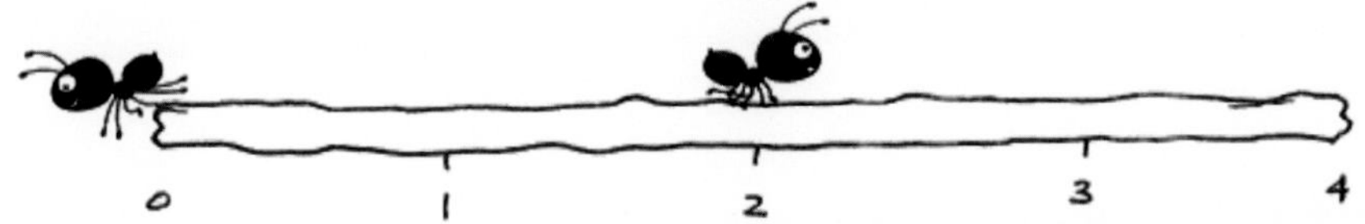

and the right one would walk
three metres right.
That would take 3/4 of a minute.
In total, one minute would have passed.

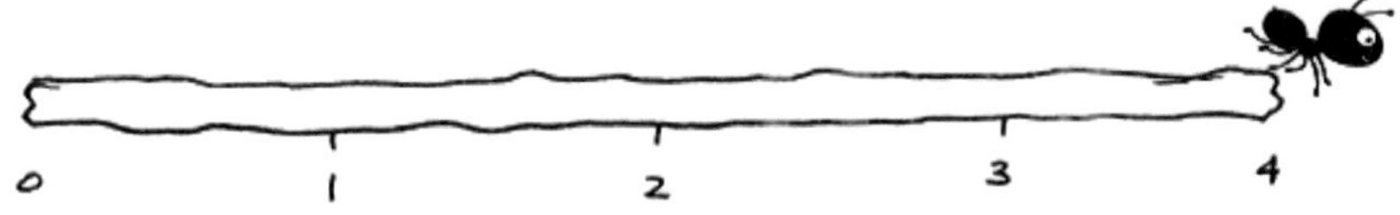

Suppose they started
somewhere along the stick...
I am quite sure
it would take not more than
one minute for both ants
to jump off!"

The boy was feeling more and more excited.
"I have a sneaky suspicion
that it would be at most one minute
even if there were seven ants!" he exclaimed.

"Let me check my theory with three ants,
to see if it is true,"
the boy said confidently.
And the boy tried one arrangement
of three ants,
and then another,
and then another...

And the boy was confused.
There were too many arrangements
of three ants...
"I can't follow the three ants!
I give up!"

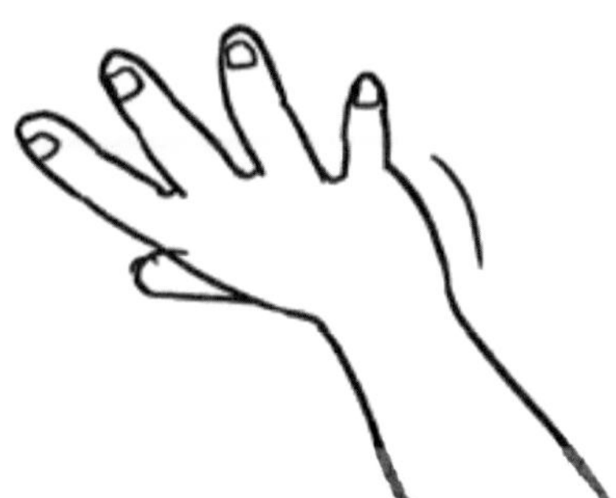

The river gurgled cheerfully
and the sun hid behind the clouds.

The boy looked up and said,
"I need a different point of view.

Let me be even more imaginative.

Let's have some fun!

Imagine a coloured cap on each ant.

Red, orange, yellow, green,
blue, indigo and violet
—the colours of the rainbow!

Seven caps for seven ants.
Now watch them go.

Imagine a stick that's four metres long with
seven colour-capped ants on it!
Back and forth they move.
Hello! Goodbye! Turn and go!

But this time
the ants will do something more.

When two meet, they will exchange caps.

Hello! Exchange caps!
Bye! Turn and go!

Look at the moving caps.
Don't look at the moving ants.

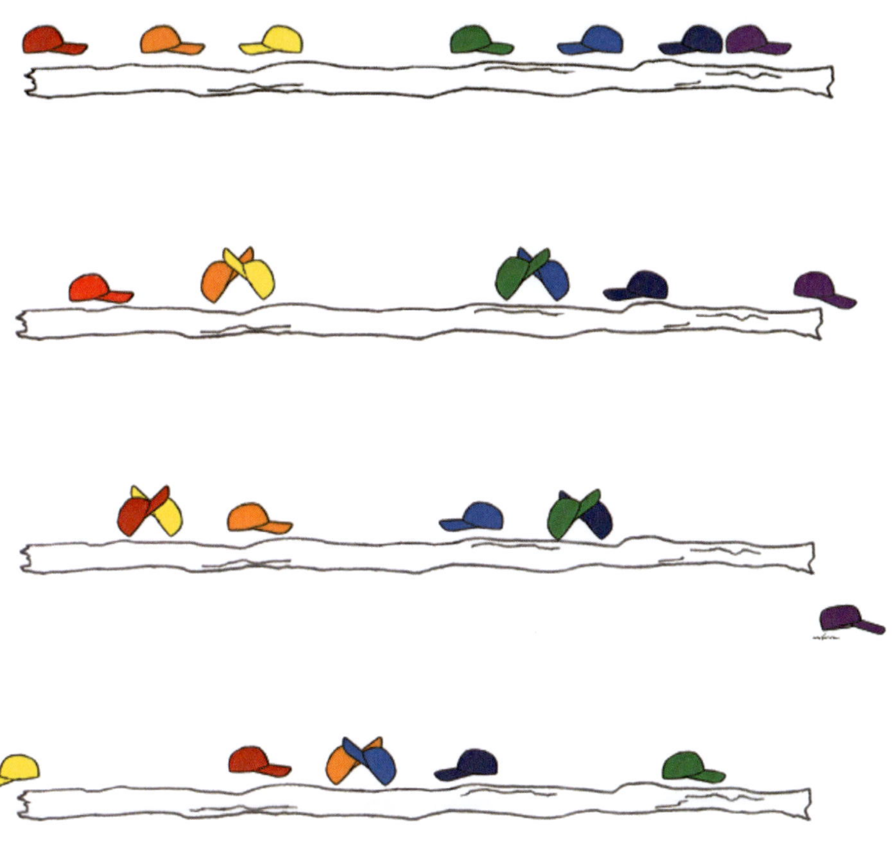

The river gurgled cheerfully
and the wind blew gently.

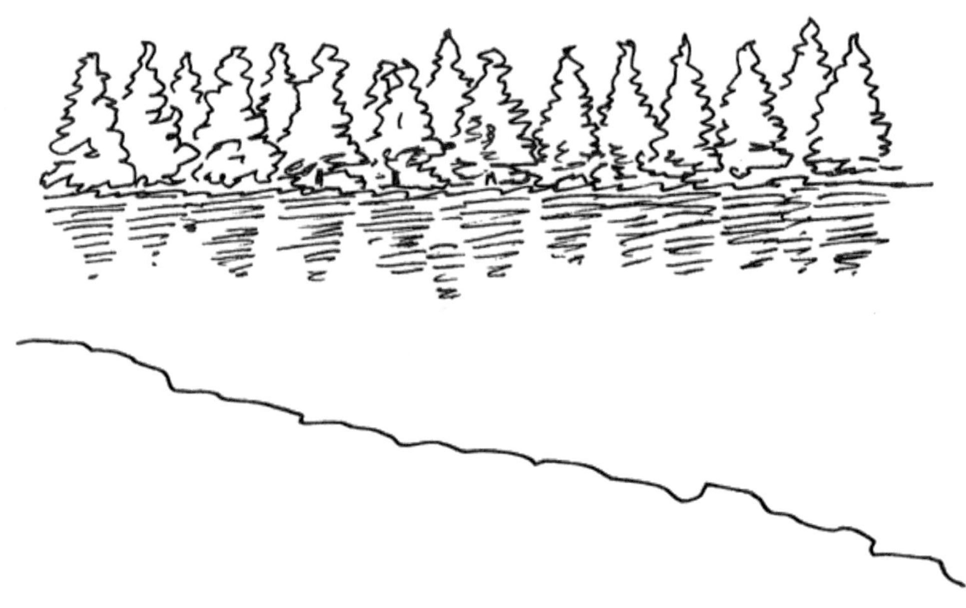

The boy smiled.

A little imagination.

A change of view.

Follow the red cap.
The red cap moves straight ahead

without changing direction.

The furthest a cap moves
is from end to end
That takes one minute.
The caps fall off the edge
The ants do the same.

And the boy was happy.

The End

The *I'm a Maths Star!* Guide to Problem-solving
Yeap Ban Har

What can we do when faced with a challenging Mathematics problem such as the one in *The Boy and the Ants*? How can we prepare ourselves to be good at problem-solving?

Here are seven tips to keep in mind:

Tip #1:
Be like the boy: look for Maths in the world around us
"The boy laid contently on the grass and saw some ants on a stick."

- The boy put on a "mathematical lens" while watching the ants moving this way and that way.

- Problem-posing is a great activity that allows us to see the structure of Maths problems.

- **Look around you and see what Maths problems you can find to solve.**

- When we are good at this, we can recognise new problems as being similar to the ones we have solved before.

Be like the boy: be curious and ask questions

The boy was fascinated. "How long would it take for all the ants to jump off?" he wondered.

- **Asking questions** that are relevant to a given situation suggests that we are able to see the important information in that situation.

- **Answering questions** that we ask along the way when we solve challenging problems often gives us information that are useful to solve those problems.

Be like the boy: visualise the situation

"Let me imagine something simpler."

- Visualisation is the ability to see using our mind, without using our eyes.

- We can **draw pictures and diagrams, including bar models** to help us visualise.

Tip #4:
Be like the boy: solve simpler problems
"If there were only one ant …."

- The boy started with one ant on the stick. Then he tried with two ants on the stick. The original problem was about seven ants on the stick.

- **Simplify a problem** by changing the numbers.

- We can understand the problem better as we solve the simpler problems. We may also be able to gain interesting and useful insights.

Tip #5:
Be like the boy: make conjectures
"I have a sneaky suspicion that it would be at most one minute even if there were seven ants!" he exclaimed.

- The boy also made sure he proved his conjecture.

- Sometimes, we may have a feeling that something might be possible. **Follow your instincts** but remember, don't jump to any conclusion without making sure that it actually works.

Be like the boy: take a step back when stuck

"I need a different point of view..."

- The boy was confused because there were many possible arrangements for the case of three ants. He was inspired to use the coloured caps representation which made things clearer.

- When we are stuck, relax. **Try to approach or represent the problem in a different way.**

Be like the boy: find joy in solving problems

- **Even if we don't arrive at a solution right away, it is fine.** The process itself makes us a better thinker.

- When we take the time to **celebrate solving problems**, we will gain motivation to solve further problems we encounter.

For parents and teachers

Pólya's Model of Problem-solving

We would suggest that any attempt at mathematical problem-solving requires a model to which the problem solver can refer, especially when one is unable to progress satisfactorily. A model helps in regulating a problem-solving attempt. We shall now describe the essential features of the problem-solving model proposed by George Pólya, an eminent Hungarian mathematician. The process can be summarised in the diagram below.

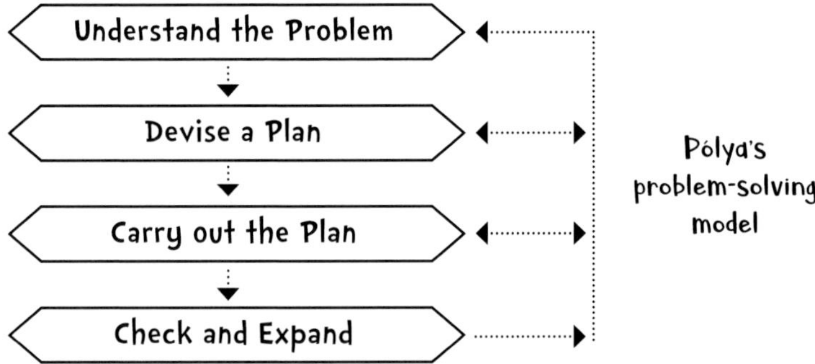

The model resembles a flowchart with four stages, Understand the Problem, Devise a Plan, Carry out the Plan, and Check and Expand, which are hierarchical but which allow back-flow. We shall explain the stages of the model and how the model works by applying it to our story of *The Boy and The Ants*.

Understand the Problem

Read aloud with the child. Pause to admire the nice illustrations. Take time to discuss and answer the child's questions, especially with regard to the number of ants and their movements. The key parts of the problem to take time to understand would be:

- Ants move at a steady pace of four metres per minute even when changing direction
- Ants fall off the edge of the stick

It is common to see people staring blankly when they are stuck at a problem. It is important that something is done and the following describes how heuristics (meaning: serving to find out or discover, *Oxford English Dictionary*) can help start the problem-solving process.

Two useful heuristics to help understand the problem are to act it out and to consider a simpler problem (smaller numbers). Consider the problem when there are only one or two ants. Then one can use simple tokens to represent the one ant or two ants and act out the scenarios as in the story. Greater clarity will result from this worthwhile use of time.

Devise a Plan / Carry out the Plan

A plan is vital to the success of any endeavour. It has been said, "He who fails to plan, plans to fail." In the story, the boy planned to proceed with three ants as he did with one or two ants. However, he found that had become too complicated. He then made a decision to change his plan. This resulted in a beautiful representation of the problem as one of moving coloured caps rather than of bouncing ants!

Check and Expand

It is good to look back at the solution and check it again. Check for various easy combinations of three or four ants to make sure that one minute is never exceeded.

We are done for this problem but not done with the problem-solving process. It is a key feature of the model that the solver should try to 'expand' the problem even though it is solved. By expanding, we mean one of the following:
- understand the underlying structure of the solution
- pose new problems

Some people have imagined a solution where the ants merely charge right through each other, exchanging bodies along the way, to understand that the movement of each ant is essentially just straight ahead.

We pose the following problems for the reader's consideration.
- What if the length of the stick were longer? Shorter?
- What if the speed of the ants moving along the stick were faster? Slower?
- What if the ants slowed down by one second every time they turned around?

By thinking through the new problems, the child will reinforce his/her understanding of the original problem.

Scan this QR code to get additional practice on how to solve similar problems. Solutions are provided too.

Learn to solve difficult Maths problems using Maths heuristics, as identified by the Singapore Maths curriculum!

The *I'm a Maths Star!* series comprises challenging Maths puzzles, presented in story form. Puzzles are described and then solved, step-by-step, through an engaging storyline, and using various Maths heuristic techniques that are also taught as part of the world-renowned Singapore Maths curriculum. Through illustrations and an engaging storyline, children will learn Maths heuristics, and be inspired to persevere in understanding, representing, and solving fun and intriguing Maths problems!

To receive updates about children's titles from WS Education, go to: *https://www.worldscientific.com/page/newsletter/subscribe*, choose "Education", click on "Children's Books" and key in your email address.

Follow us @worldscientificedu on Instagram and @World Scientific Education on YouTube for our latest releases, videos and promotions.